William Schaus, W. G Clements

On a Collection of Sierra Leone Lepidoptera

William Schaus, W. G Clements

On a Collection of Sierra Leone Lepidoptera

ISBN/EAN: 9783743323827

Manufactured in Europe, USA, Canada, Australia, Japa

Cover: Foto ©berggeist007 / pixelio.de

Manufactured and distributed by brebook publishing software
(www.brebook.com)

William Schaus, W. G Clements

On a Collection of Sierra Leone Lepidoptera

INTRODUCTION.

THE Lepidoptera which are catalogued and described in the following pages were collected by me during a tour of service of thirteen months' duration at Sierra Leone, West Africa, in the years 1891–92.

With few exceptions their habitat was the rocky peninsula of Sierra Leone, which has an area of some twenty square miles. It is very hilly, some of the hills rising to a height of from 3000 to 4000 feet. The immediately surrounding country is flat and swampy, and, so far as I had an opportunity of judging, is poor in Lepidoptera. Owing probably to its elevation, Sierra Leone is very rich in representatives of the genus *Charaxes*. In the limited time at my disposal I obtained eighteen species—nineteen, including *Palla* (*Charaxes*) *varanes*—with both sexes in fifteen of these.

In consequence of my having lost the notes I made at the time on the natural habits of the insects collected, the observations which I am enabled to make must necessarily be brief, fragmentary, and quoted from memory. They may be divided roughly into two classes: the first comprising those which have remained in aristocratic solitude, and are essentially sylvan in their habits; the second class, those which seem to prefer the vicinity of man and his habitations. Of this latter the more noticeable are the Zygænidæ amongst the Heterocera, and *D. chrysippus*, various species of *Myca-lesis* and *Precis*, *H. dædalus*, *H. misippus* (female com-paratively rare), *Terias* sp., and *P. demoleus*, *menestheus*, *phorcas*, *leonidas*, and *pylades*. The more distinctly West-African genera (*Pseudacræa*, *Euryphene*, *Euphædra*, *Aterica*, *Cymothoë*, *Epitola*) and the Ethiopian ones (*Lachnoptera*, *Salamis*, *Euxanthe*, and *Charaxes*) confine themselves to the woods and forests.

The butterflies of the genus *Aterica* fly sedately along shaded paths and among the trees, keeping near to the ground and frequently settling. They are gregarious insects, and three or four are fond of resting together, when, but for the presence of the female *opis*, which is conspicuously banded with yellow, they would be difficult to distinguish.

The *Euphædræ* fly swiftly through the lower branches of the trees. A gleam of sunlight piercing the foliage lights them up for an instant and then they disappear, having settled either on the upper surface of a leaf or on the leaves covering the ground.

The members of the genus *Euryphene* have similar habits to those of *Aterica*, but their flight is stronger. A variety of the female, which is larger and has the upper surface of its hind wings of a golden-green colour, has the figures " 1881 " very distinctly marked in black on the superior surface of the secondaries, "18" being in one and "81" in the other discoidal cell.

The manner of flight of the genus *Epitola* is similar to that of *Iolaus*. They keep flying round in small but gradually widening circles in open patches in forests, keeping some thirty or more feet from the ground ; they select the most sunlit side of a tree, on the leaves of which they frequently rest and bask in the sun.

A female of *Papilio antimachus* was brought to me in December 1891. It had been caught by hand, by a child, in a garden of a village a few miles from Freetown whilst settling on a flower. It was in perfect condition, but, to my surprise, was smaller than the male. A short description of a female *antimachus* from the Gaboon River appears in the 'Entomological Monthly Magazine' for June 1892, p. 162. A very fine male was caught in June of that year by Lieut. Stevens, of the West India Regiment, in front of the Military Sanitarium, on Kortright Hill, which is about a thousand feet above sea-level. It had settled on a stone, and flew away on being approached; curiously enough it returned to the same spot on the following day and was then captured. This insect is of rare occurrence at Sierra Leone. The females of many of the butterflies are either rare, or, what is more probable, keep themselves in retirement. The male of *P. phorcas* abounded everywhere, but I only obtained two specimens of the female, and these came from the forests clothing the less frequented hills.

The year is divided into two well-marked seasons at Sierra Leone, the wet and the dry. During the former from 170 inches to over 200 inches of rain may fall, and but few Lepidoptera are met with on the wing. At its conclusion, in November, species new to one who has only resided there during the preceding six or seven months make their appearance; even during the dry season, November to May, butterflies do not seem to occur in any great profusion. The Pierinæ are but sparsely represented, and there are none of those migratory flights of butterflies which are noted as occurring in Ceylon and the West Indies. The most striking of this family to the eye is *Tachyris chloris*; it is gregarious, often selects the vicinity of running water, over which it circles in a wavy, graceful flight; its colours of white, orange, and sooty black forming a notable contrast to the dark green foliage and sparkling water. Another butterfly which is very pleasing to watch is the long-tailed *Hypolycæna lebona* as it flies with an easy deliberation through the closest thickets.

Of the females of *P. merope* I obtained but one variety. This is indistinguishable on the wing from *Amauris niavius*; the male keeps within the shady parts of the forest and has a bold, lofty, and sailing flight. Selecting an open glade it will fly rapidly up and down for a space of some 300 yards, coming fearlessly near to one's net, which it generally manages to evade by a quick double, and finally, approaching a tuft of grass or the projecting branch of a bush, disappears. It almost invariably selects a broad-bladed grass, striped with brown and yellow, and, hanging pendent from its extremity with the wings folded, the upper ones being covered over and concealed by the lower, it cannot be seen until it is again startled into flight. Another butterfly, in this case the female, which selects a resting-place which effectually conceals it, is *Catopsilia florella*; this yellow insect has small round silvery spots, surrounded by a narrow brown margin on its wings. When near a mango-tree, of which some few of the leaves are of a bright yellow colour, dotted with spots identical in colour, shape, and disposition with those above described, it invariably selects these leaves for settling on, and is then very difficult to detect.

In bringing these few notes to a conclusion I would wish to express my great indebtedness to Mr. William Schaus, F.Z.S., for his very kind help and assistance in " working up" the Lepidoptera which I brought back from Sierra Leone. The descriptions of new species are from his pen. Mr. H. H. Druce was also so kind as to look over and name some of the Lycænidæ. Other new species, including a few Rhopalocera, will, we hope, be described shortly.

W. G. CLEMENTS,
Surgeon-Captain A.M.S.

February 1893.

ON A COLLECTION

OF

SIERRA LEONE LEPIDOPTERA.

Fam. NYMPHALIDÆ.

Subfam. DANAINÆ.

Danais chrysippus.

Papilio chrysippus, Linn. Mus. Ulr. p. 263.

Amauris damocles.

Papilio damocles, Beauv. Ins. Afr. Am. p. 239, t. 6. f. 3 *a, b.*

Amauris hecate.

Danais hecate, Butl. Proc. Zool. Soc. Lond. 1866, p. 44.

Amauris niavius.

Papilio niavius, Linn. Mus. Ulr. p. 253.

Subfam. SATYRINÆ.

Gnophodes parmeno.

Gnophodes parmeno, Doubl. & Hew. Gen. D. L. p. 363, t. 61. f. 2.

Gnophodes morpena.

Gnophodes morpena, Butl. Cat. Sat. B. M. p. 7, n. 2.

Melanitis leda.

Papilio leda, Linn. Syst. Nat. i. 2. p. 773.

Mycalesis saga.

Mycalesis saga, Butl. Cat. Sat. B. M. p. 130, t. 3. f. 1.

Mycalesis vulgaris.

Mycalesis vulgaris, Butl. Cat. Sat. B. M. p. 130, t. 3. f. 2.

Mycalesis tænias.

Mycalesis tænias, Hew. Ex. Butt. v. *Myc.* & *Idiom.* f. 66.

Ypthima itonia.

Ypthima itonia, Hew. Trans. Ent. Soc. ser. iii. vol. ii. p. 287, t. 18. f. 13.

Ypthima dolecta.

Ypthima dolecta, Kirby, Proc. Roy. Dubl. Soc. (2) ii. p. 336.

Subfam. ELYMNIINÆ.

Elymnias bammakoo.

Elymnias bammakoo, Westw. Gen. D. L. p. 405, n. 12, note.

Subfam. ACRÆINÆ.

Telchinia eponina.

Papilio eponina, Cram. Pap. Ex. iii. t. 268, A, B.

Telchinia bonasia.

Papilio bonasia, Fabr. Syst. Ent. p. 464.

Acræa quirina.

Papilio quirina, Fabr. Spec. Ins. ii. p. 36.

Acræa lycia.

Papilio lycia, Fabr. Syst. Ent. p. 461.

Acræa camœna.

Acræa camœna, Dru. Ill. Ex. Ent. ii. t. 7. f. 2.

Acræa egina.
Papilio egina, Cram. Pap. Ex. i. t. 39, F, G.

Acræa pseudegina.
Acræa pseudegina, Westw. Gen. D. L. p. 531.

Acræa orina.
Acræa orina, Hew. Ex. Butt. v. *Acr.* t. 7. ff. 43, 48.

Acræa peneleos.
Acræa peneleos, Ward, Ent. Mo. Mag. viii. p. 60.

Acræa pentapolis.
Acræa pentapolis, Ward, Ent. Mo. Mag. viii. p. 69.

Acræa circeis.
Papilio circeis, Dru. Ill. Ex. Ent. iii. t. 18. ff. 5, 6.

Acræa zetes.
Papilio zetes, Linn. Syst. Nat. i. 2. p. 766.

Acræa salambo.
Acræa salambo, Smith, Ann. Nat. Hist. (5) xix. p. 62.

Planema lycoa.
Acræa lycoa, Godt. Enc. Méth. ix. p. 239.

Planema euryta.
Papilio euryta, Linn. Mus. Ulr. p. 221.

Planema umbra.
Papilio umbra, Dru. Ill. Ex. Ent. iii. t. 18. ff. 1, 2.

Planema gea.
Papilio gea, Fabr. Spec. Ins. ii. p. 32.

Subfam. NYMPHALINÆ.

Atella columbina.
Papilio columbina, Cram. Pap. Ex. iii. t. 238, A, B.

Lachnoptera iole.
Papilio iole, Fabr. Spec. Ins. ii. p. 78.

Hypanartia delius.
Papilio delius, Dru. Ill. Ex. Ent. iii. t. 14. ff. 5, 6.

Pyrameis cardui.
Papilio cardui, Linn. Faun. Suec. p. 276.

Junonia clelia.
Papilio clelia, Cram. Pap. Ex. i. t. 21, E, F.

Junonia cebrene.
Junonia cebrene, Trim. Trans. Ent. Soc. 1870, p. 353.

Precis sinuata.
Precis sinuata, Plötz, Stett. ent. Zeit. xli. p. 477.

Precis octavia.
Papilio octavia, Cram. Pap. Ex. ii. t. 135, B, C.

Precis amestris.
Papilio amestris, Dru. Ill. Ex. Ent. iii. t. 20. ff. 3, 4.

Precis terea.
Papilio terea, Dru. Ill. Ex. Ent. ii. t. 18. ff. 3, 4.

Precis sophia.
Papilio sophia, Fabr. Ent. Syst. iii. 1. p. 248.

Precis orthosia.
Precis orthosia, Klug, Symb. Phys. t. 48. ff. 8, 9.

Precis cloantha.
Papilio cloantha, Cram. Pap. Ex. iv. t. 338, A, B.

Salamis cytora.
Salamis cytora, Doubl. & Hew. Gen. D. L. t. 25. f. 5.

Salamis anacardii.
Papilio anacardii, Linn. Mus. Ulr. p. 236.

Eurytela dryope.
Papilio dryope, Cram. Pap. Ex. i. t. 78, E, F.

Eurytela ethosea.
  Papilio ethosea, Dru. Ill. Ex. Ent. iii. t. 37. ff. 3, 4.

Hypanis ilithyia.
  Papilio ilithyia, Dru. Ill. Ex. Ent. ii. t. 17. ff. 1, 2.

Cyrestis camillus.
  Papilio camillus, Fabr. Spec. Ins. ii. p. 11.

Hypolimnas misippus.
  Papilio misippus, Linn. Mus. Ulr. p. 264.

Hypolimnas salmacis.
  Papilio salmacis, Dru. Ill. Ex. Ent. ii. t. 8. ff. 1, 2.

Hypolimnas dubius.
  Papilio dubius, Beauv. Ins. Afr. Amér. p. 238, t. 6.
    f. 2 *a, b.*

Hypolimnas anthedon.
  Diadema anthedon, Doubl. Ann. Nat. Hist. xvi. p. 181
    (1845).

Godartia eurinome.
  Papilio eurinome, Cram. Pap. Ex. i. t. 70, A.

Pseudacræa semire.
  Papilio semire, Cram. Pap. Ex. iii. t. 194, B, C.

Pseudacræa lucretia.
  Papilio lucretia, Cram. Pap. Ex. i. t. 45, C, D.

Pseudacræa dolomena.
  Diadema dolomena, Hew. Ex. Butt. iii. *Diad.* t. 2. f. 4.

Pseudacræa boisduvalii.
  Diadema boisduvalii, Doubl. Ann. Nat. Hist. xvi. p. 180
    (1845).

Pseudacræa epigea.
  Pseudacræa epigea, Butl. Cist. Ent. i. p. 214.

Catuna crithea.
  Papilio crithea, Dru. Ill. Ex. Ent. ii. t. 16. ff. 5, 6.

Catuna cœnobita.
   Hesperia cœnobita, Fabr. Ent. Syst. iii. 1. p. 247.

Neptis melicerta.
   Papilio melicerta, Dru. Ill. Ex. Ent. ii. t. 19. ff. 3, 4.

Neptis agatha.
   Papilio agatha, Cram. Pap. Ex. iv. t. 327, A, B.

Neptis nemetes.
   Neptis nemetes, Hew. Ex. Butt. iv. *Nept.* t. 1. ff. 1, 2.

Euryphene sophus.
   Papilio sophus, Fabr. Ent. Syst. iii. 1. p. 46.

Euryphene plautilla.
   Euryphene plautilla, Hew. Ex. Butt. iii. *Eur.* t. 3. ff. 14, 15.

Euryphene phantasia.
   Euryphene phantasia, Hew. Ex. Butt. iii. *Eur.* t. 2. ff. 9–11.

Euryphene sœmis.
   Euryphene sœmis, Hew. Ex. Butt. iii. *Eur.* t. 1. ff. 1, 2.

Euryphene tentyris.
   Euryphene tentyris, Hew. Ex. Butt. iii. *Eur.* t. 5. ff. 21, 22.

Euryphene gambiæ.
   Euryphene gambiæ, Feisth. Ann. Soc. Ent. Fr. 1850, p. 251,
      t. 9. f. 2.

Euryphene brunhilda.
   Euryphene brunhilda, Kirby, Ann. Nat. Hist. (6) iii. p. 247.

Euphædra perseis.
   Papilio perseis, Dru. Ill. Ex. Ent. ii. t. 21. ff. 3, 4.

Euphædra zampa.
   Romaleosoma zampa, Westw. Gen. D. L. p. 284.

Euphædra ceres.
   Papilio ceres, Fabr. Syst. Ent. p. 504.

Euphædra janassa.

Papilio janassa, Linn. Mus. Ulr. p. 294.

Euphædra francina.

Nymphalis francina, Godt. Enc. Méth. ix. p. 390, n. 141.

Euphædra harpalyce.

Papilio harpalyce, Cram. Pap. Ex. ii. t. 145, D, E.

Euphædra xypete.

Romaleosoma xypete, Hew. Ex. Butt. iii. *Rom.* t. 2. ff. 8–10.

Euphædra inanum.

Romaleosoma inanum, Butl. Cist. Ent. i. p. 158.

Euphædra arcadius.

Papilio arcadius, Fabr. Ent. Syst. iii. 1. p. 151.

Euphædra agnes.

Romaleosoma agnes, Butl. Proc. Zool. Soc. Lond. 1865, p. 672.

Hamanumida dædalus.

Papilio dædalus, Fabr. Syst. Ent. p. 482.

Aterica ampedusa.

Aterica ampedusa, Hew. Ex. Butt. iii. *At. & Euryphene*, t. 5. ff. 3–5.

Aterica veronica.

Papilio veronica, Cram. Pap. Ex. iv. t. 325, C, D.

Aterica cupavia.

Papilio cupavia, Cram. Pap. Ex. iii. t. 193, E, F.

Aterica opis.

Papilio opis, Dru. Ill. Ex. Ent. ii. t. 18. ff. 5, 6.

Cymothoë cænis.

Papilio cænis, Dru. Ill. Ex. Ent. ii. t. 19. ff. 1, 2.

Cymothoë sangaris.

Nymphalis sangaris, Godt. Enc. Méth. ix. p. 384.

Cymothoë egesta.

Papilio egesta, Cram. Pap. Ex. i. t. 46, B, C.

Cymothoë usilda.

Harma usilda, Hew. Ex. Butt. iv. *Har*. t. 4. ff. 13, 14.

Charaxes epijasius.

Charaxes epijasius, Reiche, Ferr. Gal. Voy. Abyss., Ent. p. 469, t. 32. ff. 1, 2.

Charaxes pollux.

Papilio pollux, Cram. Pap. Ex. i. t. 37, E, F.

Charaxes castor.

Papilio castor, Cram. Pap. Ex. i. t. 37, C, D.

Charaxes brutus.

Papilio brutus, Cram. Pap. Ex. iii. t. 211, E, F.

Charaxes lucretius.

Papilio lucretius, Cram. Pap. Ex. i. t. 82, E, F.

Charaxes protoclea.

Charaxes protoclea, Feisth. Ann. Soc. Ent. Fr. 1850, p. 260.

Charaxes anticlea.

Papilio anticlea, Dru. Ill. Ex. Ent. iii. t. 27. ff. 5, 6.

Charaxes candiope.

Nymphalis candiope, Godt. Enc. Méth. ix. p. 353, n. 10.

Charaxes etesipe.

Nymphalis etesipe, Godt. Enc. Méth. ix. p. 355, n. 19.

Charaxes tiridates.

Papilio tiridates, Cram. Pap. Ex. ii. t. 161, A, B.

Charaxes numenes.

Charaxes numenes, Hew. Ex. Butt. ii. *Nymph*. t. 2. ff. 9–11.

Charaxes ameliæ.

   Charaxes ameliæ, Doum. Rev. Zool. 1861, p. 171, t. 5.
     f. 1.

Charaxes mycerina.

   Charaxes mycerina, Godt. Enc. Méth. ix. p. 369, n. 65.

Charaxes smaragdalis.

   Charaxes smaragdalis, Butl. Proc. Zool. Soc. Lond. 1865,
     p. 630, t. 36. f. 5.

Charaxes imperialis.

   Charaxes imperialis, Murr. Trans. Ent. Soc. 1874,
     p. 531.

Charaxes alladinis.

   Charaxes alladinis, Butl. Cist. Ent. i. p. 5.

Charaxes eupale.

   Papilio eupale, Dru. Ill. Ex. Ent. iii. t. 6. f. 3.

Charaxes zingha.

   Papilio zingha, Cram. Pap. Ex. iii. t. 315, B, C (1782).

Palla ussheri.

   Palla ussheri, Butl. Trans. Ent. Soc. 1870, p. 124.

Palla (Charaxes) varanes.

   Papilio varanes, Cram. Pap. Ex. ii. t. 160, D, E.

Palla lichas.

   Palla lichas, Doubl. & Hew. Gen. D. L. t. 49. f. 3.

Subfam. **LIBYTHÆINÆ.**

Libythea labdaca.

   Libythea labdaca, Westw. Gen. D. L. p. 413, t. 68. f. 6.

Fam. LYCÆNIDÆ.

Epitola urania.

Epitola urania, Kirby, Ann. & Mag. Nat. Hist. (5) xix. p. 441.

Epitola teresa.

Epitola teresa, Hew. Ent. Mo. Mag. vi. p. 86.

Epitola dewitzi.

Epitola dewitzi, Kirby, Ann. Nat. Hist. (5) xix. p. 442.

Epitola badura.

Epitola badura, Kirby, Ann. Nat. Hist. (6) iv. p. 271.

Epitola versicolor.

Epitola versicolor, Kirby, Ann. Nat. Hist. (5) xix. p. 444.

Epitola, sp.?

Epitola, sp.?, near *Pinodes*, H. H. Druce.

Argiolaus ælianus.

Argiolaus ælianus, Stg. Iris, iv. p. 148.

Argiolaus belli.

Argiolaus belli, Hew. Ann. Nat. Hist. (4) xiii. p. 382.

Argiolaus laon.

Argiolaus laon, Hew. Ill. Diurn. Lep., Supp. p. 28, pl. iv. ff. 46, 47.

Castalius isis.

Papilio isis, Dru. Ill. Ex. Ent. ii. t. 3. ff. 4, 5.

Lycænesthes larydas.

Papilio larydas, Cram. Pap. Ex. iii. t. 282, H.

Lycænesthes lysicles.

Lycænesthes lysicles, Hew. Trans. Ent. Soc. 1874, p. 348.

Lycænesthes moncus.

Papilio moncus, Fabr. Spec. Ins. ii. p. 113.

Lycænesthes amarah.

Polyommatus amarah, Guér. Lef. Voy. Abyss. vi. p. 384, t. 11. ff. 5, 6.

Lycænesthes sichela.

Lycæna sichela, Wallengr. Lep. Rhop. Caffr. p. 37.

Lycæna cissus.

Polyommatus cissus, Godt. Enc. Méth. ix. p. 683.

Lycæna juba.

Papilio juba, Fabr. Mant. Ins. p. 82.

Lycæna lingeus.

Papilio lingeus, Cram. Pap. Ex. iv. t. 379, F, G (1782).

Lycæna parsimon.

Papilio parsimon, Fabr. Syst. Ent. p. 526.

Lycæna telicanus.

Papilio telicanus, Hübn. Eur. Schmett. i. ff. 371, 372, 553, 554.

Lyæna lysimon.

Lycæna lysimon, Hübn. Eur. Schmett. i. ff. 534, 535.

Lycæna knysna.

Lycæna knysna, Trimen, Trans. Ent. Soc. ser. iii. vol. i. p. 282.

Lycæna cyclopteris.

Lamprospilus cyclopteris, Butl. Ann. Nat. Hist. (4) xviii. p. 483.

Lycæna docilis.

Lycæna docilis, Butl. Proc. Zool. Soc. Lond. 1887, p. 571.

Cyaniris micylus.

Papilio micylus, Cram. Pap. Ex. iii. t. 282, F, G.

Epamera iaspis.

Epamera iaspis, H. H. Druce, Ann. Mag. Nat. Hist. (6) vol. v. p. 50.

Plebeius gussfeldtii.

> *Plebeius gussfeldtii*, Dewitz, Nova Acta, B. xli. p. 11, no. 2, p. 206, t. xxvii. f. 12.

Larinopoda muhata.

> *Larinopoda muhata*, Smith & Kirby, Rhop. Ex. *Lyc.* pl. ii. ff. 1–4.

Tingra preussi.

> *Tingra preussi*, Staud. Ex. Schmett. p. 267.

Axiocerces perion.

> *Papilio perion*, Cram. Pap. Ex. iv. t. 379, B, C.

Zeritis fallax.

> *Zeritis fallax*, Sharpe, Ann. Mag. Nat. Hist., July 1890, p. 104.

Zeritis latifimbriata.

> *Zeritis latifimbriata*, Sharpe, Ann. Mag. Nat. Hist., July 1890, p. 105.

Zeritis bicolor.

> *Zeritis bicolor*, Sharpe, Ann. Mag. Nat. Hist., Sept. 1891, p. 241.

Myrina silenus.

> *Papilio silenus*, Fabr. Syst. Ent. p. 531, n. 378.

Myrina genuba.

> *Myrina genuba*, Hew. Ent. Mo. Mag. xii. p. 106.

Myrina nomion.

> *Myrina nomion*, Stg. Iris, iv. p. 156, pl. i. f. 11.

Spindasis clymenus.

> *Spindasis clymenus*, H. H. Druce, Ent. Mo. Mag. xvii. p. 259.

Aphnæus orcas.

> *Papilio orcas*, Dru. Ill. Ex. Ent. iii. t. 34. ff. 2, 3.

Aphnæus natalensis.

Aphnæus natalensis, Doubl. & Hew. Gen. D. L. t. 75. f. 4.

Pseudaletis trifasciata.

Pseudaletis trifasciata, Sharpe, Ann. Nat. Hist., July 1890, p. 103; Trans. Ent. Soc. p. 556, pl. xviii. f. 8.

Deudorix galathea.

Thecla galathea, Swains. Zool. Ill. ii. t. 69.

Deudorix deritas.

Deudorix deritas, Hew. Trans. Ent. Soc. 1874, p. 352.

Deudorix anta.

Lycena anta, Trim. Trans. Eut. Soc. ser. iii. vol. i. p. 402.

Iolaus eurisus.

Iolaus eurisus, Cram. Pap. Ex. iii. t. 221, D, E.

Iolaus iasis.

Iolaus iasis, Hew. Ill. D. L. p. 42, t. 19. ff. 11, 12.

Iolaus iulus.

Iolaus iulus, Hew. Ill. D. L. Suppl. p. 9, t. 4. ff. 41–43.

Iolaus calisto.

Anthene calisto, Doubl. & Hew. Gen. D. L. t. 75. f. 6.

Iolaus timon.

Papilio timon, Fabr. Mant. Ins. ii. p. 65.

Hypolycæna philippus.

Hesperia philippus, Fabr. Ent. Syst. iii. 1. p. 283.

Hypolycæna lebona.

Hypolycæna lebona, Hew. Ill. D. L. p. 51, n. 9, t. 23. ff. 28, 29.

Hypolycæna antifaunus.

Iolaus antifaunus, Doubl. & Hew. Gen. D. L. t. 75. f. 1.

Hypolycæna hatita.

Hypolycæna hatita, Hew. Ill. D. L. p. 51, t. 23. ff. 21–24.

Hypolycæna eleala.

Hypolycæna eleala, Hew. Ill. D. L. p. 52, t. 23. ff. 25–27.

Hypolycæna hymen.

Papilio hymen, Fabr. Syst. Ent. p. 519.

Hypolycæna mera.

Hypolycæna mera, Hew. Ent. Mo. Mag. x. p. 124.

Hypolycæna naara.

Hypolycæna naara, Hew. Ent. Mo. Mag. x. p. 124.

Hypolycæna gracilis.

Hypolycæna gracilis, Stg. Iris, iv. p. 152, pl. i. f. 9.

Hypolycæna zela.

Hypolycæna zela, Hew. Ill. D. L. Suppl. p. 14, n. 22, t. 5. ff. 41–43.

Pseudodipsas dewitzii.

Pseudodipsas dewitzii, Stg. Iris, iv. p. 155, pl. i. f. 10.

Fam. PAPILIONIDÆ.

Subfam. PIERINÆ.

Nychitona alcesta.

Papilio alcesta, Cram. Pap. Ex. iv. t. 379, A.

Terias zoe.

Terias zoe, Hopff. Ber. Verh. Ak. Berl. 1855, p. 640, n. 5.

Terias senegalensis.

Terias senegalensis, Boisd. Sp. Gén. i. p. 672.

Pieris calypso.

Papilio calypso, Dru. Ill. Ex. Ent. ii. t. 17. ff. 3, 4.

Pieris hedyle.

Papilio hedyle, Cram. Pap. Ex. ii. t. 186, C, D.

Tachyris chloris.
Papilio chloris, Fabr. Syst. Ent. p. 473, n. 129.

Tachyris rhodope.
Papilio rhodope, Fabr. Syst. Ent. p. 473, n. 130.

Tachyris sylvia.
Papilio sylvia, Fabr. Syst. Ent. p. 470, n. 115.

Eronea poppea.
Papilio poppea, Don. Nat. Rep. ii. t. 54. f. 2.

Eronia pharis.
Pieris pharis, Boisd. Sp. Gén. i. p. 443.

Eronia argia.
Papilio argia, Fabr. Syst. Ent. p. 470.

Catopsilia florella.
Papilio florella, Fabr. Syst. Ent. p. 479.

Teracolus arethusa.
Papilio arethusa, Dru. Ill. Ex. Ent. ii. t. 19. ff. 5, 6.

Subfam. **PAPILIONINÆ**.

Papilio antimachus.
Papilio antimachus, Dru. Ill. Ex. Ent. iii. t. 1.
A female in perfect condition.

Papilio leonidas.
Papilio leonidas, Fabr. Syst. Ent. iii. 1. p. 35.

Papilio tynderæus.
Papilio tynderæus, Fabr. Syst. Ent. iii. 1. p. 35.

Papilio latreillanus.
Papilio latreillanus, Godt. Enc. Méth. ix. p. 44.

Papilio menestheus.
Papilio menestheus, Dru. Ill. Ex. Ent. ii. t. 9. ff. 1, 2.

Papilio demoleus.
  Papilio demoleus, Linn. Mus. Ulr. p. 214.

Papilio antheus.
  Papilio antheus, Cram. Pap. Ex. iii. t. 234, B, C.

Papilio pylades.
  Papilio pylades, Fabr. Syst. Ent. iii. 1. p. 34.

Papilio bromius.
  Papilio bromius, Doubl. Ann. Nat. Hist. xvi. p. 176 (1845).

Papilio nireus.
  Papilio nireus, Linn. Mus. Ulr. p. 217.

Papilio phorcas.
  Papilio phorcas, Cram. Pap. Ex. i. t. 2, B, C.

Papilio merope.
  Papilio merope, Cram. Pap. Ex. ii. t. 151, A, B.

Papilio cynorta.
  Papilio cynorta, Fabr. Syst. Ent. iii. 1. p. 37.

Papilio zenobia.
  Papilio zenobia, Fabr. Syst. Ent. p. 503. n. 255.

Papilio cypræafila.
  Papilio cypræafila, Butl. Ent. Mo. Mag. v. p. 60.

Fam. HESPERIDÆ.

Ismene forestan.
  Papilio forestan, Cram. Pap. Ex. iv. t. 391, E, F.

Ismene pisistratus.
  Hesperia pisistratus, Fabr. Syst. Ent. iii. 1. p. 345.

Ismene iphis.
  Papilio iphis, Dru. Ill. Ex. Ent. ii. t. 15. ff. 3, 4.

Ismene bixæ.

Papilio bixæ, Linn. Mus. Ulr. p. 335.

Pamphila philander.

Carystus philander, Hopff. Ber. Verh. Ak. Berl. 1855,
p. 643.

Proteides helops.

Papilio helops, Dru. Ill. Ex. Ent. iii. t. 33. ff. 2, 3.

Pamphila borbonica.

Hesperia borbonica, Boisd. Faun. Mad. p. 65, n. 3, t. 9.
ff. 5, 6.

Pamphila ignita.

Pamphila ignita, Mab. Bull. Soc. Ent. Fr. (5) vii. pl. xl.

Ceratrichia phocion.

Papilio phocion, Fabr. Spec. Ius. ii. p. 138.

Pardaleodes edipus.

Papilio edipus, Cram. Pap. Ex. iv. t. 366, E, F.

Pardaleodes laronia.

Hesperia laronia, Hew. Desc. Hesp. p. 35.

Fam. SPHINGIDÆ.

Subfam. MACROGLOSSINÆ.

Cephonodes hylas.

Sphinx hylas, Linn. Mant. Plant. p. 539.

Macroglossa trochiloides.

Macroglossa trochiloides, Butl. Proc. Zool. Soc. Lond.
1875, p. 5.

Macroglossa falkensteinii.

Macroglossa falkensteinii, Dewitz, Mitth. Münch. Ent. Ver.
iii. p. 23, t. 1. f. 1.

Lophura pylas.

   Sphinx pylas, Cram. Pap. Exot. iii. t. 206, A.

Diodosida iapygoides.

   Ocyton iapygoides, Holl. Trans. Amer. Ent. Soc. xvi. p. 60,
      t. 2. f. 5.

Antinephele anomala.

   Nephele anomala, Butl. Ann. Nat. Hist. (5) x. p. 434.

Subfam. CHŒROCAMPINÆ.

Basiothia medea.

   Sphinx medea, Fabr. Spec. Ins. ii. p. 143, n. 19.

Chœrocampa eson.

   Sphinx eson, Cram. Pap. Exot. iii. t. 226, C.

Chœrocampa gracilis.

   Chœrocampa gracilis, Butl. Proc. Zool. Soc. Lond. 1875,
      p. 8, t. 2. f. 2.

Chœrocampa charis.

   Chœrocampa charis, Walk. Cat. Lep. Het. B. M. viii.
      p. 136, n. 15.

Chœrocampa osiris.

   Sphinx osiris, Dalm. Anal. Ent. p. 48, n. 21.

Chœrocampa balsaminæ.

   Chœrocampa balsaminæ, Walk. Cat. Lep. Het. B. M. viii.
      p. 138, n. 18.

Chœrocampa clotho.

   Sphinx clotho, Dru. Ill. Ex. Ent. ii. t. 28. f. 1.

Daphnis nerii.

   Sphinx nerii, Linn. Syst. Nat. i. p. 490, n. 5.

Subfam. **AMBULICINÆ**.

Nephele æquivalens.
   Pachylia æquivalens, Walk. Cat. Lep. Het. B. M. viii.
      p. 191, n. 5.

Nephele bipartita.
   Nephele bipartita, Butl. Ann. Nat. Hist. (5) ii. p. 455.

Nephele œnopion.
   Orneus œnopion, Hübn. Samml. Ex. Schmett. ii.

Nephele variegata.
   Nephele variegata, Butl. Proc. Zool. Soc. Lond. 1875,
      p. 15.

Nephele funebris.
   Sphinx funebris, Fabr. Ent. Syst. iii. 1. p. 371, n. 47.

Subfam. **SPHINGINÆ**.

Protoparce solani.
   Sphinx solani, Boisd. Faune Madag. p. 76, t. 11. f. 2.

Protoparce convolvuli.
   Sphinx convolvuli, Linn. Syst. Nat. i. p. 490, n. 6.

Subfam. **MANDUCINÆ**.

Manduca atropos.
   Sphinx atropos, Linn. Syst. Nat. i. p. 490, n. 8.

Subfam. **SMERINTHINÆ**.

Basiana postica.
   Basiana postica, Walk. Cat. Lep. Het. B. M. viii. p. 237,
      n. 3.

Basiana stigmatica.
   Basiana stigmatica, Mab. Bull. Soc. Zool. France, ii. p. 491.

Fam. AGARISTIDÆ.

Anaphela terminatis.

Eusemia terminatis, Walk. Cat. Lep. Het. B. M. vii. p. 1587.

Xanthospilopteryx euphemia.

Noctua euphemia, Stoll, Pap. Ex. iv. t. 345, A.

Massaga maritona.

Massaga maritona, Butl. Proc. Zool. Soc. Lond. 1868, p. 224, t. 17. f. 1.

Massaga delicia.

Massaga delicia, Butl. Proc. Zool. Soc. Lond. 1868, p. 224, t. 17. f. 2.

Ægocera boisduvalii.

Ægocera boisduvalii, Latr., Cuvier, Règne Anim. (ed. 11) iii. t. 20. f. 3.

Ægocera leucomelas.

Phægorista leucomelas, Herr.-Schäff. Aussereur. Schmett. i. ff. 22, 23.

Ægocera leona, sp. nov. (Plate I. fig. 1.)

Primaries above brown; at the base a large white spot starting from the costal margin and extending to the middle of the wing above the submedian vein; just beyond the cell a small white transverse spot not reaching the costa; between these spots some transverse velvety brown shades, and a submarginal, very angular, velvety brown line: underneath dull brown, with the base broadly white, especially along the inner margin, and the white transverse spot beyond the cell larger than on the upper surface. Secondaries white, with the outer margin very broadly black, especially at the apex; underneath, the marginal band is dull brown. Body brown; palpi and fore coxæ reddish brown.

Expanse 38 millim.

Charilina amabilis.

Noctua amabilis, Dru. Ill. Ex. Ent. ii. t. 13. f. 3.

Fam. C H A L C O S I I D Æ.

Pseudopontia paradoxa.

Globiceps paradoxa, Feld. Pet. nouv. Ent. no. 8.

Fam. T H Y M A R I D Æ.

Pedoptila nemopteridia.

Pedoptila nemopteridia, Butl. Ann. Nat. Hist. (5) xv. p. 341.

Fam. Z Y G Æ N I D Æ.

Subfam. ZYGÆNINÆ.

Syntomis cerbera.

Sphinx cerbera, Linn. Mus. Ulr. p. 363.

Syntomis touvasina.

Syntomis touvasina, Butl. Journ. Linn. Soc. Lond., Zool. xii. p. 348.

Syntomis divalis, sp. nov.   (Plate I. fig. 2.)

Primaries above black; at the base a large crimson spot; about the middle three hyaline spots, the largest being in the cell, and the next below it being the smallest; the anterior and posterior hyaline spots are followed by a round crimson spot, beyond which are three smaller hyaline spots followed by a subapical crimson band. Secondaries above black, with the costal margin yellowish. Underneath, the wings are dull black, with the base of the primaries and the costal margin of the secondaries yellowish. Antennæ black, with white tips. Frons yellow. Collar black, with two yellow spots. Thorax black, with a dorsal reddish line and a lateral red spot. Abdomen above crimson, with a black transverse band on each segment. Underneath, the body is yellowish, the legs outwardly streaked with brown.

Expanse 34 millim.

Subfam. THYRETINÆ.

Thyretes caffra.

Thyretes caffra, Wallengr. Wien. Ent. Mon. vii. p. 138.

Eressa bivittata.

Syntomis bivittata, Walk. Cat. Lep. Het. B. M. xxxi. p. 66.

Saluinca aurifrons.

Saluinca aurifrons, Walk. Cat. Lep. Het. B. M. xxxi. p. 109.

Subfam. EUCHROMIINÆ.

Euchromia leonis.

Euchromia leonis, Butl. Journ. Linn. Soc. Lond., Zool. xii. p. 383.

Euchromia fulvida.

Euchromia fulvida, Butl. Trans. Ent. Soc. Lond. 1888, p. 112, t. 4. f. 5.

Larva feeds upon sweet potato (*Batatas edulis*). It has been noticed feeding at mid-day, in the full glare of the sun. All the species of this family are fond of flying about garden-hedges, preferably those of *Lantana*. They are to be found sparingly all the year round, but appear in greater numbers in August, towards the cessation of the heavy rains. Period of pupation about 12 days. (*W. G. C.*)

Fam. ARCTIIDÆ.

Subfam. CHARIDEINÆ.

Charidea vicaria.

Euchromia vicaria, Walk. Cat. Lep. Het. B. M. i. p. 207.

Subfam. **PHÆGOPTERINÆ.**

Anace burra, sp. nov.   (Plate I. fig. 6.)

Primaries and body pinkish fawn-colour.   Secondaries paler, with only the extreme margins darker.   Antennæ dark brown; tarsi brown.

Expanse 31 millim.

Metarctia erubescens.

Metarctia erubescens, Walk. Cat. Lep. Het. B. M. xxxi. p. 315.

Casphalia picta, sp. nov.   (Plate II. fig. 6.)

Primaries with the basal third yellow, except a small black space at the base itself; the entire costa and outer portion of the wing black; a subapical oblique yellow spot.   Secondaries orange; the outer margin broadly black; a small round black spot beyond the cell.   Head and collar orange.   Thorax brown.   Abdomen orange; the anal segment black.

Expanse 32 millim.

Subfam. **SPILOSOMATINÆ.**

Alpenus maculosus.

Bombyx maculosa, Stoll, Pap. Ex. iv. t. 370, B.

Fam. LITHOSIIDÆ.

Utetheisa pulchella.

Tinea pulchella, Linn. Syst. Nat. i. p. 534.

Argina leonina.

Deiopeia leonina, Walk. Cat. Lep. Het. B. M. xxxi. p. 262.

Argina cingulifera?

Deiopeia cingulifera, Walk. Cat. Lep. Het. B. M. ii. p. 569.

Fam. HYPSIDÆ.

Pseudhypsa speciosa.

Noctua speciosa, Dru. Ill. Ex. Ent. ii. t. 5. f. 2.

Caryatis syntomina.

Caryatis syntomina, Butl. Ann. Nat. Hist. (5) ii. p. 456.

Godasa maculatrix.

Godasa maculatrix, Walk. Cat. Lep. Het. B. M. xxxi. p. 271.

Fam. NYCTEMERIDÆ.

Aletis flammea, sp. nov. (Plate II. fig. 2.)

Primaries above and below black, crossed just beyond the middle by an oblique, broad white band, from the costal vein to nearly the inner angle. Secondaries fiery red, yellowish along the costal margin ; the base shaded with blackish scales, and a broad black outer margin, becoming quite narrow at the anal angle. Head red ; a black spot on the frons. Collar red, with two black spots. Tegulæ red, with a black streak. Abdomen above red ; a subdorsal row of black spots : underneath yellow, with a lateral row of black spots. Legs brown outwardly, yellowish inwardly ; fore coxæ red, with a small black spot.

Expanse 50 millim.

Pitthea continua.

Pitthea continua, Walk. Cat. Lep. Het. B. M. ii. p. 463.

Nyctemera perspicua.

Nyctemera perspicua, Walk. Cat. Lep. Het. B. M. ii. p. 398.

Nyctemera apicalis.

Nyctemera apicalis, Walk. Cat. Lep. Het. B. M. ii. p. 395.

Amnemopsyche famula.

Bombyx famula, Dru. Ill. Ex. Ent. ii. t. 11. f. 3.

Amnemopsyche gracilis.

Amnemopsyche gracilis, Möschl. Abhandl. Senckenb. Ges.
xv. p. 73. f. 1.

Otrœda occidentis.

Otrœda occidentis, Walk. Cat. Lep. Het. B. M. ii. p. 403.

Otrœda nerina.

Bombyx nerina, Dru. Ill. Ex. Ent. iii. t. 5. f. 1.

Fam. LIPARIDÆ.

Geodena quadrigutta.

Geodena quadrigutta, Walk. Cat. Lep. Het. B. M. vii.
p. 1691.

Eloria divisa.

Eloria divisa, Walk. Cat. Lep. Het. B. M. iv. p. 815.

Redoa laba, sp. nov. (Plate I. fig. 4.)

Entirely snow-white, except the palpi, frons, tarsi, and fore
coxæ, which are orange. In the female the costal margin of
the primaries is narrowly yellow.

Expanse, ♂ 33 millim., ♀ 42 millim.

Antiphella crocicollis.

Liparis crocicollis, Herr.-Schäff. Aussereur. Schmett. i.
f. 110.

Cypra eleuteria.

Bombyx eleuteria, Stoll, Suppl. Cram. t. 36. f. 13.

Crorema mentiens.

Crorema mentiens, Walk. Cat. Lep. Het. B. M. iv. p. 811.

Artaxa modesta, sp. nov. (Plate II. fig. 4.)

Entirely pale yellow.

Expanse 25 millim.

Chrysopsyche mirifica.

Chærotriche mirifica, Butl. Ann. Nat. Hist. (5) ii. p. 458.

Utidava? citana, sp. nov. (Plate I. fig. 10.)

Primaries above pale buff; the costal margin to nearly the apex very broadly blackish brown, this colour occupying nearly the entire cell, and enclosing at the end of the cell a cluster of much darker scales; the inner margin broadly shaded with light brown; the outer margin from the apex broadly blackish brown, this shade being inwardly dentate and crossed by a marginal row of small buff spots, outwardly edged with blackish scales. Secondaries above yellowish white. Underneath yellowish white, the costal margin of the secondaries and the entire primaries shaded with brown. Head and thorax dark brown. Abdomen lighter brown.

Expanse 24 millim.

Lælia fracta, sp. nov. (Plate I. fig. 12.)

Primaries above light buff, finely speckled with dark scales; a dark shade extending from the base, through the cell, to almost the outer margin; on the posterior two thirds of the wing a marginal oblique row of small dark spots; the fringe alternately buff and grey. Secondaries whitish. Underneath, the wings are dirty white. Body greyish.

Expanse 21 millim.

Lælia rosea, sp. nov. (Plate I. fig. 5.)

Primaries above pink, thinly speckled with dark scales; a curved marginal row of black points. Secondaries pinkish white. Body pinkish.

Expanse 22 millim.

Aroa danva, sp. nov. (Plate I. fig. 3.)

Male. Primaries above brown, darker along the costal margin, beyond the cell, and at the apex, except a light shade on the costal margin at about two thirds from the base; some greenish scales at the base of the costal vein, below the subcostal vein, and at the inner angle; an inner, outer, and submarginal transverse irregular dark line, in part very indistinct; the fringe alternately light and dark brown: underneath yellow; at the base of the inner margin a broad brownish space; the apical third of the wing brownish, faintly mottled with yellow. Secondaries above brown, darkest along the costal margin; an orange band starting near the base, and gradually widening towards the middle of the outer margin: underneath yellowish, shaded with brown; a large dark brown space at the apex; a transverse wavy brown line not reaching the inner margin, and a dark lunular mark in the cell. Body above dark brown, underneath light brown.

Expanse 39 millim.

Orgyia ticana, sp. nov. (Plate I. fig. 11.)

Primaries above dark brown; along the inner margin and at the apex yellowish; a basal, median, and outer transverse angular dark line, partly bordered with yellowish where approaching the lighter shades on the wing; at the end of the cell two short dark transverse streaks: underneath light brown, darker in the disc; an outer transverse nearly straight brown line. Secondaries above greyish brown: underneath light brown; a small brown ring in the cell and two transverse brownish lines, the outer one being indistinct and broken. Body brown.

Expanse 20 millim.

Dasychira acrisia.

Deiopeia (?) *acrisia*, Plötz, Stett. ent. Zeit. xli. p. 83.

Fam. LIMACODIDÆ.

Heterolepis plötzi, sp. nov. (Plate II. fig. 1.)

Primaries above with the costal margin and outer third of the wing light brown; at the apex a terminal velvety brown line, preceded by a broad but short dark brown mark; otherwise violaceous brown, with a black transverse basal streak, and beyond this a more conspicuous transverse wavy black line from the cell to the inner margin; at the end of the cell a black spot. Secondaries above greyish brown; all the fringes very long, yellowish brown. Underneath, the wings are yellowish brown, sprinkled with darker scales; the outer margins indistinctly outlined with dark scales. Body light brown; the anterior portion of the thorax, two clusters of scales dorsally at the base of the abdomen, and the fore coxæ dark velvety brown.

Expanse 20 millim.

Allied to *Heterolepis leprosa*, Felder.

Miresa syrtis, sp. nov.   (Plate II. fig. 3.)

Primaries above brown, thinly speckled with darker scales, especially on the veins; a dark transverse line, slightly curved, from the costal margin at two thirds from the base to the middle of the inner margin; a similar, nearly straight submarginal line. Secondaries brown, with a few dark scales on the outer margin near the apex. Underneath, the wings are uniformly brown. Body brown.

Expanse 19 millim.

Allied to *Miresa hilda*, Druce.

Fam. NOTODONTIDÆ.

Rigema ornata.

Rigema ornata, Walk. Cat. Lep. Het. B. M. xxxii. p. 437.

Anthena simplex.

Anthena simplex, Walk. Cat. Lep. Het. B. M. iii. p. 687.

Fam. BOMBYCIDÆ.

Naroma signifera.

Naroma signifera, Walk. Cat. Lep. Het. B. M. vii. p. 1744.

Fam. DREPANULIDÆ.

Plegapteryx anomalus.

Plegapteryx anomalus, Herr.-Schäff. Aussereur. Schmett. i. ff. 462, 463.

Fam. SATURNIIDÆ.

Bunæa eblis.

Bunæa eblis, Streck. Lep. p. 121, t. 14. f. 9.

Bunæa jamesoni.

Bunæa jamesoni, Druce, Jameson, Story of Rear Column, p. 448.

Imbrasia obscura.

Gonimbrasia obscura, Butl. Ann. Nat. Hist. (5) ii. p. 462.

Antheræa arata.

Saturnia arata, Westw. Proc. Zool. Soc. Lond. 1849, p. 41, t. 7. f. 2.

Antheræa dione.

Bombyx dione, Fabr. Ent. Syst. iii. (1) p. 410.

Eudæmonia brachyura.

Attacus brachyura, Dru. Ill. Ex. Ent. iii. t. 29. f. 1.

Eudæmonia argiphontes.

Eudæmonia argiphontes, Kirby, Trans. Ent. Soc. Lond. 1877, p. 20.

Bolocera smilax.

Saturnia smilax, Westw. Proc. Zool. Soc. Lond. 1849, p. 59, n. 31.

Fam. LASIOCAMPIDÆ.

Stibolepis cunina.

Attacus cunina, Cram. Pap. Exot. iii. t. 237, G.

Stibolepis odites, sp. nov.   (Plate I. fig. 9.)

Wings above snowy white; on the primaries the outer portion of the veins finely outlined with brown. Underneath, the costal and extreme outer margins, also the veins on the secondaries, yellowish. Head and collar orange. Thorax white. Abdomen yellowish, with indistinct transverse greyish shades.

Expanse 48 millim.

Homochroa? orphne, sp. nov.   (Plate I. fig. 7.)

Male. Primaries above greyish white; some velvety brown marks at the base; at the end of the cell, which is very short, an oblique black spot containing a whitish line; two median, outwardly curved, lunular transverse brownish shades, followed by other similar but more indistinct shades; the apex brownish, forming a dark patch on the outer margin below the apex, and an oblique dark shade on the costal margin; fringe brown, with longitudinal white streaks. Secondaries above yellowish white; the veins finely brown towards the outer margin; some long orange scales towards the base; some velvety-brown markings along the inner margin; two transverse angular brown lines from the costal to the inner margin. Underneath creamy white; the veins and extreme outer margin brown; the costal margin of the primaries orange, and on the same wings a dark shade below the apex on the outer margin; on the secondaries the markings of the upper surface are repeated, but the transverse lines are followed by some submarginal lunular marks. Head and collar brown. Thorax mottled brown and white. Abdomen dorsally brownish, laterally white, underneath orange.

Expanse 62 millim.

Jana eurymas.

Jana eurymas, Herr.-Schäff. Aussereur. Schmett. f. 98.

Jana strigina.

Lasiocampa strigina, Westw. Proc. Zool. Soc. Lond. 1849,
p. 37.

Lichenopteryx fulvia.

Lichenopteryx fulvia, Druce, Proc. Zool. Soc. Lond. 1886,
p. 411.

Stenoglene rudis.

Lasiocampa rudis, Walk. Cat. Lep. Het. B. M. xxxii.
p. 561.

Dendrolimus bipars.

Lebeda bipars, Walk. Cat. Lep. Het. B. M. vi. p. 1455.

Dendrolimus ferruginea.

Pachypasa ferruginea, Feld. Reise d. Novara, t. 85. f. 1.

Dendrolimus basalis.

Megasoma basale, Walk. Cat. Lep. Het. B. M. vi. p. 1448.

Gastropacha gerstaeckeri.

Gastropacha gerstaeckeri, Dewitz, Verh. Leop.-Car. Akad.
xlii. p. 74, t. 1. f. 6.

Lasiocampa heres, sp. nov. (Plate I. fig. 8.)|

Primaries above olivaceous, shaded with brown, the middle
of the inner margin being the palest ; a basal and an outer
transverse darker line, the former slightly undulating, the
latter forming an angle beyond the cell and reaching the
inner margin in its centre. Secondaries olivaceous brown,
paler on the costal margin ; an ill-defined blackish marginal
shade ; a transverse line starting from the costal margin and
only noticeable on the lighter portion of the wing. Underneath olivaceous brown, with a transverse undulating darker
shade, chiefly noticeable on the secondaries. Body olivaceous brown ; the head and palpi brown.

Expanse 52 millim.

Lasiocampa mæra. (Plate II. fig. 5.)

Primaries above greenish yellow, shaded with reddish brown along the costal margin and below the cell; a broad submarginal transverse shade of a dull brown, this shade being very irregular exteriorly; a basal and a median transverse reddish-brown line, and a small spot at the end of the cell of a similar colour. Secondaries above dark brown, shaded with reddish along the costal margin. Underneath, the wings are reddish brown, with the inner margin of the secondaries broadly dark brown. Head, thorax, and anal segment yellowish green. Abdomen brown; lighter underneath.

Expanse 44 millim.

Fam. PINARIDÆ.

Cyrtogone herilia.

Saturnia herilia, Westw. Proc. Zool. Soc. Lond. 1849, p. 57, t. 11. f. 3.

Cyrtogone nenia.

Saturnia nenia, Westw. Proc. Zool. Soc. Lond. 1849, p. 57, t. 9. f. 3.

Gonometa cassandra.

Gonometa cassandra, Druce, Proc. Zool. Soc. Lond. 1887, p. 681.

Gonometa matuta, sp. nov. (Plate II. fig. 8.)

Female. Primaries above greyish brown; a black point at the end of the cell; two median transverse wavy grey lines; a submarginal irregular transverse greyish shade. Secondaries yellowish; the costal margin greyish brown. Underneath, all the wings brownish yellow. Head and thorax greyish brown. Abdomen yellowish.

Expanse 78 millim.

Fam. ZEUZERIDÆ.

Zeuzera boisduvalii.

Zeuzera boisduvalii, Herr.-Schäff. Aussereur. Schmett. i. f. 167.

Cossus toluminus.

Cossus toluminus, Druce, Proc. Zool. Soc. Lond. 1887, p. 684.

Fam. NOCTUIDÆ.

Opigena accipiter, sp. nov.   (Plate II. fig. 7.)

Primaries above brown; the basal third blackish, crossed near the base and outwardly shaded by two transverse greyish lines; some greyish and velvety brown scales at the end of the cell; an irregular submarginal dark shade, on either side of which are some greyish scales. Secondaries light brown; the veins somewhat darker; an indistinct transverse whitish shade. Underneath, the wings are dirty brown, with a median and a submarginal transverse dark shade; a dark spot on the middle of the costal margin of the primaries, and a still larger spot on the costal margin of the secondaries nearer the base.   Body dark grey.

Expanse 43 millim.

Ochropleura talda, sp. nov.   (Plate III. fig. 1.)

Primaries dark grey, lighter on the outer margin; the orbicular and reniform very indistinct; between the two some velvety black scales, and beyond the reniform some similar scales; along the outer margin some dark longitudinal streaks between the veins. Secondaries white; the fringe at the apex smoky.   Body grey; two dark brown spots on the collar.

Expanse 25 millim.

Perigea africana, sp. nov.   (Plate II. fig. 10.)

Primaries brown; at the base of the costa a white spot and another below it; just before the orbicular a white

wavy mark on the costa, and a round white spot below the
median vein; the orbicular surrounded by four white points;
the reniform fawn-colour, surrounded and broken by white
spots; a white spot on the costa above the reniform; between
this and a subapical white spot a few white points on the
costa; a white spot near the middle of the outer margin, and
a similar spot on the inner margin near the angle; beyond
the reniform some indistinct transverse lines.   Secondaries
brownish.   Body brown.

Expanse 30 millim.

Very similar to *Perigea stelligera,* Guenée.

Spodoptera mauritia.

Hadena mauritia, Boisd. Faun. Ent. Mad. 92, 3. pl. 13.
f. 9.

Glottula pancratii.

Noctua pancratii, Cyrillo, Ent. Neapol. pl. 12. f. 4.

Euthisanotia florifera.

Polytela florifera, Walk. Cat. Lep. Het. B. M. xv. p. 1666.

Meliana bertha, sp. nov.   (Plate III. fig. 3.)

Primaries above buff, thickly speckled with black scales;
beyond the cell a conspicuous black spot, preceded by a
much smaller one, and followed by a transverse row of black
points; on the extreme margin a row of black points; a dark
shade from the large black spot to the apex: underneath
whitish buff, with a black spot on the costal margin at four
fifths from the base.   Secondaries whitish, slightly smoky
on the outer margin.   Body buff.

Expanse 25 millim.

Curubasa lanceolata.

Maria lanceolata, Walk. Cat. Lep. Het. B. M. xxxiii.
p. 767.

Episparis penetrata.

Episparis penetrata, Walk. Cat. Lep. Het. B. M. x. p. 476.

Xanthodes graellsii.

Acontia graellsii, Feisth. Ann. Soc. Ent. Fr. vi. p. 300,
pl. 12. f. 3.

Acontia camilla.

Xanthodes camilla, Druce, Proc. Zool. Soc. Lond. 1887,
p. 686.

Leocyma polla, sp. nov. (Plate II. fig. 11.)

Primaries silvery whitish ; a broad irregular median transverse brownish band, and a submarginal narrow brown shade,
widening at the costal margin ; the extreme margin finely
brown ; fringe greyish. Secondaries light brown. Head
and collar yellow. Thorax and abdomen whitish brown.

Expanse 22 millim.

Leocyma pollusca, sp. nov. (Plate II. fig. 12.)

Primaries light brown, slightly tinged with violaceous ;
two fine and straight transverse brown lines, the first oblique,
the second beyond the middle of the wing. Secondaries dark
brown ; the extreme margin paler. Body brown.

Expanse 27 millim.

Leocyma fustina, sp. nov. (Plate II. fig. 9.)

Primaries above rich brown ; the orbicular represented
by a white point ; the reniform yellowish, containing a white
point ; on the posterior half of the wing a submarginal broad
yellowish shade ; on the extreme outer margin a row of black
points outwardly shaded with yellowish. Secondaries dull
brown ; the fringe lighter. Underneath light brown, with an
outer transverse brownish line ; black points on the extreme
outer margins ; a black discal spot on the secondaries.

Expanse 38 millim.

Earias fervida.

Earias fervida, Walk. Cat. Lep. Het. B. M. xxxv. p. 1774.

Xanthoptera colla, sp. nov. (Plate III. fig. 6.)

Primaries light grey ; the costal margin paler ; at the apex

a small black mark inwardly shaded with white scales; the
fringe yellowish. Secondaries yellowish white. Underneath,
the wings are yellowish, the disc of the primaries shaded
with brown. Head and collar orange. Thorax and abdomen
whitish grey.

Expanse 18 millim.

Xanthoptera allecta, sp. nov. (Plate III. fig. 4.)

Primaries with the basal half yellow and the outer half
violaceous brown; at the base of the inner margin a blackish
spot and a basal transverse wavy line; in the cell a black
point, above which, on the costal margin, is a brown spot;
the middle third of the outer half occupied by a transverse
wavy line of lighter brown. Secondaries dull brown; the
fringe reddish. Head and thorax yellow; the abdomen
brownish.

Expanse 23 millim.

Tarache pyralina.

Acontia pyralina, Walk. Cat. Lep. Het. B. M. xii. p. 789.

Tarache perta, sp. nov.   (Plate III. fig. 2.)

Primaries above snowy white; at the base some brownish
marks on the costa; a broad median transverse brown band
widest on the inner margin; the outer margin broadly shaded
with brown, especially towards the apex, this shade advancing
inwardly towards the cell. Secondaries grey. Head and
thorax white; abdomen yellowish grey.

Expanse 15 millim.

Gonitis fulvida.

Anomis fulvida, Guen. Noct. ii. p. 397.

Gonitis leona, sp. nov.   (Plate III. fig. 10.)

Primaries above brown, tinged with violaceous; the apex
slightly paler; the basal line distinct, inwardly shaded with
greyish scales near the costal margin; the outer line exteriorly shaded with dark brown; a conspicuous yellow spot
below the median vein at about its middle; the fringe
blackish, with its extremity between the veins shaded with

yellowish white. Secondaries dull brown. Underneath, the primaries have the disc blackish, the costal and outer margins pale, and the apex reddish. The secondaries are reddish, with an outer wavy line.

Expanse 35 millim.

Birtha talusina, sp. nov. (Plate III. fig. 12.)

Primaries with the base dark violaceous brown from the costal margin at one fourth to the middle of the inner margin, this space being crossed by a light buff line, which forms a sharp angle on the subcostal vein; beyond this the wing is pale buff, except a large triangular dark space on the costal margin just beyond the middle, and a smaller dark space on the same margin before the apex; a small velvety spot in the cell, in an oblique line with a smaller spot on the costal margin; beyond the cell a small dark space; a wavy median transverse line and an outer straight line to the inner angle. Secondaries light brown. Body brown.

Expanse 30 millim.

Allied to *B. insulata*, Walk.

Aedia discistriga.

Anophia discistriga, Walk. Cat. Lep. Het. B. M. xiii. p. 1128.

Piala basipunctum.

Piala basipunctum, Walk. Cat. Lep. Het. B. M. xv. p. 1766.

Bareia incidens.

Bareia incidens, Walk. Cat. Lep. Het. B. M. xv. p. 1840.

Ophideres fullonica.

Phalæna-Noctua fullonica, Linn. Syst. Nat. p. 812.

Argadesa materna.

Phalæna-Noctua materna, Linn. Syst. Nat. ii. p. 840.

Halustria intricatus.

Halustria intricatus, Butler.

Trisula magnifica, sp. nov. (Plate III. fig. 8.)

Primaries light grey, mottled with pale reddish brown below the apex from the costal to the outer margin, close to the inner angle, and on the reniform; a broad velvety black basal transverse band; at two thirds from the base on the costal margin a triangular black spot, from which starts an outer irregular and indistinct line. Secondaries white, with the outer margin broadly black. Underneath, the primaries are dull brown, except a white space occupying two thirds of the inner margin, and a transverse broad whitish band; a discal spot on the secondaries below. Head and thorax reddish grey. Abdomen whitish; a quadrate black dorsal spot at the base.

Expanse 65 millim.

Miniodes discolor.

Miniodes discolor, Guen. Noct. iii. p. 119.

Patula macrops.

Phalæna-Noctua macrops, Linn. Syst. Nat. p. 225.

Cyligramma limacina.

Cyligramma limacina, Guér. Règn. Anim., Ins. pl. 89. f. 2.

Cyligramma fluctuosa.

Phalæna fluctuosa, Dru. Ill. Ex. Ent. ii. pl. 14. f. 1.

Maxula capensis.

Hypopyra capensis, Herr.-Schäff. Aussereur. Schmett. ff. 121, 122.

Ophiodes finifascia.

Nephelodes finifascia, Walk. Cat. Lep. Het. B. M. xv. p. 1676.

Ericeia inangulata.

Hulodes inangulata, Guen. Noct. iii. p. 210.

Ophisma klugii.

Ophiusa klugii, Boisd. Faun. Ent. Mad. p. 103.

Ophisma illustrata.

Achæa illustrata, Walk. Cat. Lep. Het. B. M. xiv. p. 1392.

Ophisma lienardi.

Ophiusa lienardi, Boisd. Faun. Ent. Mad. p. 102, pl. 15. f. 5.

Ophisma exhibens.

Ophisma exhibens, Walk. Cat. Lep. Het. B. M. xiv. p. 1388.

Ophisma mormoides.

Achæa mormoides, Walk. Cat. Lep. Het. B. M. xiv. p. 1393.

Achæa leona.

Achæa leona, Feld. Reise d. Nov. pl. cxvi. f. 13.

Ophiusa allardi.

Ophiusa allardi, Oberth. Etud. d'Ent. iii. p. 35, t. 2. f. 6.

Ophiusa angularis.

Ophiusa angularis, Boisd. Faun. Ent. Mad. p. 103, t. 13. f. 2.

Trigonodes hyppasia.

Phalæna-Noctua hyppasia, Cram. Pap. Exot. iii. t. 250. f. E.

Audea bipunctata.

Audea bipunctata, Walk. Cat. Lep. Het. B. M. xiii. p. 1135.

Remigia repanda.

Noctua repanda, Fabr. Ent. Syst. iii. 2. p. 49.

Remigia mayeri.

Ophiusa mayeri, Boisd. Faun. Ent. Mad. p. 104.

Remigia lituraria.

Alamis lituraria, Saalm. Lep. v. Mad. p. 419.

Focilla docta, sp. nov. (Plate III. fig. 15.)

Primaries above reddish brown; a transverse line from

near the apex to the inner margin, outwardly shaded with greyish; at the end of the cell an inwardly curved hyaline streak: underneath greyish, with a basal, median, and marginal reddish shade; at the apex a reddish-brown spot. Secondaries above reddish brown; a brighter transverse band and an irregular hyaline mark at the end of the cell: underneath reddish, with a straight transverse grey line. Body reddish brown.

Expanse 30 millim.

Azazia rubricans.

Ophiusa rubricans, Boisd. Faune Lép. Mad. p. 106, t. 16. f. 1.

Calobochyla silona, sp. nov.   (Plate III. fig. 5.)

Primaries brown, shaded with red along the costal and outer margins; from the apex to the middle of the inner margin an indistinct dull brown shade; a submarginal row of minute black spots. Secondaries dull brown, with the fringe reddish. Underneath, the wings are reddish, with the disc of the primaries brown, and the inner margin of the secondaries whitish. Body brown. Legs reddish.

Expanse 25 millim.

Hypena obacervalis.

Hypena obacervalis, Walk. Cat. Lep. Het. B. M. xvi. p. 53.

Hypena lividalis.

Pyralis lividalis, Hübn. Eur. Schmett. v. f. 11. 186.

Hypena saltalis, sp. nov.   (Plate III. fig. 14.)

Primaries above greyish brown; two velvety black spots at a third from the base, one above, the other below the median vein; an indistinct outer brown line. Secondaries greyish brown. Underneath lighter, especially along the inner margin of the secondaries. Body greyish brown.

Expanse 21 millim.

Hypena ? ducalis, sp. nov.   (Plate III. fig. 13.)

Primaries brownish, with nearly the entire basal half except the costa bright orange. Secondaries orange, with the outer margin broadly blackish. Head, collar, and anterior half of the thorax brown; posterior half of the thorax and first abdominal segments orange; the rest of the abdomen brown.

Expanse 28 millim.

Sophronia capalis.

Sophronia capalis, Walk. Cat. Lep. Het. B. M. xvi. p. 95.

Herminia mascusalis.

Herminia mascusalis, Walk. Cat. Lep. Het. B. M. xvi. p. 112.

Hydrillodes? janalis, sp. nov. (Plate III. fig. 11.)

Primaries above rich brown, crossed from the costal to the inner margin by two snowy-white irregular lines; near the apex some very indistinct marginal white scales: underneath dull brown, with the costal outer margins greyish; an indistinct outer whitish shade. Secondaries above blackish brown: underneath grey; a white discal spot and two outer wavy brown lines. Body brown above, greyish white underneath.

Expanse 20 millim.

Dragana pansalis.

Dragana pansalis, Walk. Cat. Lep. Het. B. M. xvi. p. 200.

Marimatha duplicalis.

Marimatha duplicalis, Walk. Cat. Lep. Het. B. M. xxxiv. p. 1205.

Fam. GEOMETRIDÆ.

Zamarada transvisaria.

Epione transvisaria, Guen. Uran. et Phalén. ix. p. 98.

Zamarada reflexaria.

Comibæna reflexaria, Walk. Cat. Lep. Het. B. M. xxvi. p. 1565.

Racotis squalida.

Ophthalmodes squalida, Butl. Ann. Nat. Hist. (5) ii. p. 465
 (1878).

Pingasa ruginaria.

Hypochroma ruginaria, Guen. Uran. et Phalén. ix. p. 278.

Microloxia latilineata.

Geometra latilineata, Walk. Cat. Lep. Het. B. M. xxxv.
 p. 1605.

Thalassodes congrua.

Geometra congrua, Walk. Desc. Het. Lep. from Congo,
 p. 47.

Pareumelea perlimbata.

Palyas perlimbata, Guen. Uran. et Phalén. ix. p. 396.

Rhamidava fulvata.

Phalæna fulvata, Dru. Ill. Ex. Ent. iii. t. 21. f. 4.

Rhamidava amplissimata.

Acidalia amplissimata, Walk. Cat. Lep. Het. B. M. xxvi.
 p. 1614.

Timana sodaliata.

Rhamidava sodaliata, Walk. Cat. Lep. Het. B. M. xxvi.
 p. 1568.

Anisodes leonaria.

Ephyra leonaria, Walk. Cat. Lep. Het. B. M. xxii. p. 635.

Gnamptoloma neptunaria.

Timandra neptunaria, Guen. Uran. et Phalén. x. p. 3,
 pl. 18. f. 5.

Idœa remotata.

Acidalia remotata, Guen. Uran. et Phalén. ix. p. 458.

Idœa lactaria.

Acidalia lactaria, Walk. Cat. Lep. Het. B. M. xxii. p. 744.

Idœa cleoraria.

> *Acidalia cleoraria*, Walk. Cat. Lep. Het. B. M. xxiii.
> p. 792.

Rambara melagonata.

> *Zanclopteryx melagonata*, Walk. Cat. Lep. Het. B. M. xxvi.
> p. 1619.

Acropteris erycinaria.

> *Micronia erycinaria*, Guen. Uran. et Phalén. x. p. 30.

Acropteris tenella.

> *Micronia tenella*, Walk. Desc. Het. Lep. from Congo,
> p. 53.

Dirades theclata.

> *Erosia theclata*, Guen. Uran. et Phalén. x. p. 36.

Tephrina exfusaria.

> *Aspilates exfusaria*, Walk. Cat. Lep. Het. B. M. xxvi.
> p. 1683.

Leucetœra simpliciata.

> *Cidaria simpliciata*, Walk. Cat. Lep. Het. B. M. xxv.
> p. 1422.

Fam. PYRALIDÆ.

Pachynoa phialusalis.

> *Botys phialusalis*, Walk. Cat. Lep. Het. B. M. xix. p. 991.

Cirrochrista saltusalis, sp. nov.   (Plate III. fig. 7.)

Primaries above white ; the basal third of the costa orange, this orange continuing in an oblique line to the inner margin ; beyond the cell a broad transverse orange line, posteriorly bifurcating to the middle of the inner margin and to the inner angle ; an indistinct subapical transverse orange line ;

on the extreme margin a series of black marks; the fringe orange. Secondaries silvery white; a row of black points on the extreme margin. Body white; an orange transverse band on the third abdominal segment; the anus golden, circled with black.

Expanse 24 millim.

Pharazia otreusalis.

Botys otreusalis, Walk. Cat. Lep. Het. B. M. xviii. p. 637.

Acharana cynaralis?

Botys cynaralis, Walk. Cat. Lep. Het. B. M. xviii. p. 672.

Acharana verminalis.

Botys verminalis, Guenée, Delt. et Pyral. p. 348.

Chrysotidiris hirtusalis.

Botys hirtusalis, Walk. Cat. Lep. Het. B. M. xviii. p. 642.

Pygospila tyres.

Phalæna-Pyralis tyres, Cram. Pap. Exot. iii. t. 263, C.

Margaronia baldersalis.

Margaronia baldersalis, Walk. Cat. Lep. Het. B. M. xviii. p. 527.

Margaronia ocellata.

Margaronia ocellata, Warr.

Stemorrhages sericea.

Phalæna sericea, Dru. Ill. Ex. Ent. ii. pl. 6. f. 1.

Eudioptis indica.

Eudioptis indica, Saunders, Zool. ix. p. 3070.

Tobata elealis.

Tobata elealis, Walk. Cat. Lep. Het. B. M. xviii. p. 516.

Notarcha mysisalis.

Botys mysisalis, Walk. Cat. Lep. Het. B. M. xviii. p. 634.

Haritala temeratalis.

Botys temeratalis, Zeller, Caff. p. 42.

Zebronia podalirialis.

Spilomela podalirialis, Guen. Delt. et Pyral. p. 281.

Coptobasis sarronalis.

Botys sarronalis, Walk. Cat. Lep. Het. B. M. xviii. p. 636.

Coptobasis leonalis, sp. nov.   (Plate III. fig. 9.)

Primaries brownish black ; a white spot in the cell, another above the middle of the inner margin ; beyond the cell a short row of small contiguous spots to the costal margin. Secondaries brownish black, with an indistinct whitish transverse line.   Underneath brown, with all the markings more distinct.   Body blackish above, white underneath.

Expanse 24 millim.

Hymenea recurvalis.

Phalæna recurvalis, Fabr. Ent. Syst. iii. 2. 237.

Hedylepta vulgalis.

Asopia vulgalis, Guen. Delt. et Pyral. p. 202.

Orphanostigma abruptalis.

Asopia abruptalis, Walk. Cat. Lep. Het. B. M. p. 371.

Bocchoris inspersalis.

Ædiodes inspersalis, Zeller, Caff. p. 33.

Agrotera citrina.

Agrotera citrina, Warr.

Eurrhyparodes tricoloralis.

Diasemia tricoloralis, Zeller, Caff. p. 31.

Lepyrodes geometralis.

Lepyrodes geometralis, Guen. Delt. et Pyral. p. 278.

Crochiphora testulatis.

Crochiphora testulalis, Hübn. Samml. Exot. Schmett. ff. 629, 630.

Cadarena sinuata.

Phalæna sinuata, Fabr. Ent. Syst. iii. p. 208.

Marasmia venilialis.

Asopia venilialis, Walk. Cat. Lep. Het. B. M. xvii. p. 373.

Dolichosticha trapezalis.

Salbia trapezalis, Guen. Delt. et Pyral. p. 200.

Printed by TAYLOR and FRANCIS, Red Lion Court, Fleet Street.

PLATE I.

Fig.		Page
1. Ægocera leona	. . . . .	20
2. Syntomis divalis	. . . . .	21
3. Aroa danva	. . . .	26
4. Redoa laba	. . . .	25
5. Lælia rosea	. . . .	26
6. Anace burra	. . . .	23
7. Homochroa? orpluc	. . . . .	30
8. Lasiocampa heres	. . . . . .	31
9. Stibolepis odites	. . . . .	30
10. Utidava? citana	. . . . . .	26
11. Orgyia ticana	. . . . .	27
12. Lælia fracta	. . . .	26

PLATE II.

Fig.		Page
1. Heterolepis plötzi	. . . . .	27
2. Aletis flammea	. . . . .	24
3. Miresa syrtis	. .	28
4. Artaxa modesta	. . .	26
5. Lasiocampa mæra	. . . . .	32
6. Casphalia picta	. . . .	23
7. Opigena accipiter	. . . .	33
8. Gonometa matuta	. . . .	32
9. Leocyma fustina	. . . .	35
10. Perigea africana	. .	33
11. Leocyma polla	. . . . .	35
12. Leocyma pollusca	. .	35

PLATE III.

Fig.		Page
1.	Ochropleura talda	33
2.	Tarache perta	36
3.	Meliana bertha	34
4.	Xanthoptera allecta	36
5.	Calobochyla silona	40
6.	Xanthoptera colla	35
7.	Cirrochrista saltusalis	43
8.	Trisula magnifica	38
9.	Coptobasis leonalis	45
10.	Gonitis leona	36
11.	Hydrillodes? janalis	41
12.	Birtha talusina	37
13.	Hypena? ducalis	40
14.	Hypena saltalis	40
15.	Focilla docta	39

www.ingramcontent.com/pod-product-compliance
Lightning Source LLC
Chambersburg PA
CBHW031929060726
47496CB00008BA/2777